Math for All Learners Probability and Statistics

by
Robert H. Jenkins

illustrated by
Lois Leonard Stock

User's Guide
to
Walch Reproducible Books

As part of our general effort to provide educational materials that are as practical and economical as possible, we have designated this publication a "reproducible book." The designation means that purchase of the book includes purchase of the right to limited reproduction of all pages on which this symbol appears:

Here is the basic Walch policy: We grant to individual purchasers of this book the right to make sufficient copies of reproducible pages for use by all students of a single teacher. This permission is limited to a single teacher and does not apply to entire schools or school systems, so institutions purchasing the book should pass the permission on to a single teacher. Copying of the book or its parts for resale is prohibited.

Any questions regarding this policy or requests to purchase further reproduction rights should be addressed to:

Permissions Editor
J. Weston Walch, Publisher
321 Valley Street • P. O. Box 658
Portland, Maine 04104-0658

1 2 3 4 5 6 7 8 9 10

ISBN 0-8251-4135-4

P. O. Box 658 • Portland, Maine 04104-0658
www.walch.com

Printed in the United States of America

Contents

To the Teacher

As teachers, we know that students learn best when they "know math by doing math." The activities in this book are designed to enable students to discover the concepts of probability and statistics through a hands-on approach. Many of the activities call for working in groups, to give students the support of others as they explore unfamiliar concepts; this also helps develop cooperative learning and communication skills.

Several major topics in probability and statistics are included in this book. These activities will lead the learner through the topic in an active, hands-on, minds-on approach to learning.

Each topic is addressed by a linked series of activities. Each part in the series builds on the activity before, but the multi-part structure makes it easy to present the series over the course of several classes. Each activity series includes a teacher page and one or more student activity pages, which contain:

- Student learning outcomes
- Time requirements
- Materials list
- NCTM Standards (2000) being addressed
- Prerequisites (if any)
- Overview and background information
- Procedure
- Reproducible student activity sheets

The activities in this book use several formats for cooperative learning. They include using a team format:

- to collaboratively construct meaning to a mathematical concept
- to generate oral and written responses for assessment
- to generate team and individual responses

This is accomplished through the intricate development of the activities. Different aspects of the team learning approach include:

- individual assignments, both written and oral

- team products
- individual assessment opportunities

Some of the collaborative learning structures include:

- The learning "pyramid"
- Jigsaws (regular and "reverse")
- Individual student responsibilities within the team

Grouping Strategies for Cooperative Learning

Throughout this book, grouping strategies are keyed to the activities. Teams designed to be homogeneous or heterogeneous have always been popular with teachers, but research indicates that this strategy has many inherent problems. Students tend to object to teams that are formed strategically. Randomly formed teams, on the other hands, not only tend to be more exciting but also nullify student complaints because the team assignments last for only a short span of time. These activities are mostly designed for such an experience. Therefore, the teams for all these activities are to be selected randomly and last for only one activity (unless stated otherwise).

You can use a deck of cards to form teams randomly. With the seating arranged according to the requirements of the activity, assign numbers to the work stations. Reduce the deck of cards so that there is a one-to-one correspondence with the specific class scheduled. Randomly hand out the cards as the students arrive, and have the students sit at the appropriate work stations according to the number on their card—aces sit at table 1, etc. By identifying the suit (e.g., "hearts"), individual jobs within the team can be assigned as the activity begins.

Some activities may require a jigsaw approach to learning. An easy method to assign responsibilities within a team (especially for a jigsaw activity) is to have students assign themselves numbers between 1 and 4. (Be sure to identify the number chosen by each student before continuing.) Students may then be assigned to tasks by common numbers. (See graphic organizer that follows.)

Grouping Strategies			
Jigsaw Grouping Strategy		*Pyramid Grouping Strategy*	
Step 1:	Form teams.	Step 1:	x y z
Step 2:	Member w does task 1. Member x does task 2. Member y does task 3. Member z does task 4.	Step 2:	xyz

Step 3:	Convene member *w* from each team, member *x* from each team, member *y* from each team, and member *z* from each team to share results.	Step 3:	(*x*, *y*, *z* / *a*, *b*, *c*) (*d*, *e*, *f* / *u*, *v*, *w*)
Step 4:	Each original team reconvenes.	Step 4:	(*x*, *y*, *z* *d*, *e*, *f* / *a*, *b*, *c* *u*, *v*, *w*)

Constructivist Learning Theory

The structure upon which the geometry activities in this book are designed is the constructivist learning theory. The constructivist learning cycle phases—*exploration, identification,* and *application*—are frequently used. Research and classroom practice have both produced strong support for this way of learning. Students, teachers, parents, and administrators have seen positive changes in achievement, the ability to apply learning, and the reduction of anxiety.

The *concept-exploration stage* gives students the chance to experience a concept with a hands-on, minds-on activity. Students begin by using their prior knowledge and innate abilities to construct their own knowledge about a concept within their own familiar framework. Reasoning skills, problem-solving aptitudes, and communication abilities are enhanced when learners can explore, experiment, and share. Many of the activities are designed for this stage of the learning cycle.

The *concept-identification stage* is the central part of the learning cycle. After exploring a concept, students engage in activities that synthesize and build on their learning. This stage helps the student develop a sophisticated level of mathematical literacy and master the abilities that will produce the learning outcome you're looking for.

The *concept-application stage* allows students to use their mathematical learning. Students *do* mathematics by applying their knowledge to solve problems, create products, or explore a new concept.

These three stages often overlap. Each activity allows for the diversity of student experiences, knowledge, and motivation. Many activities incorporate two or more stages of the learning cycle.

Spinnering

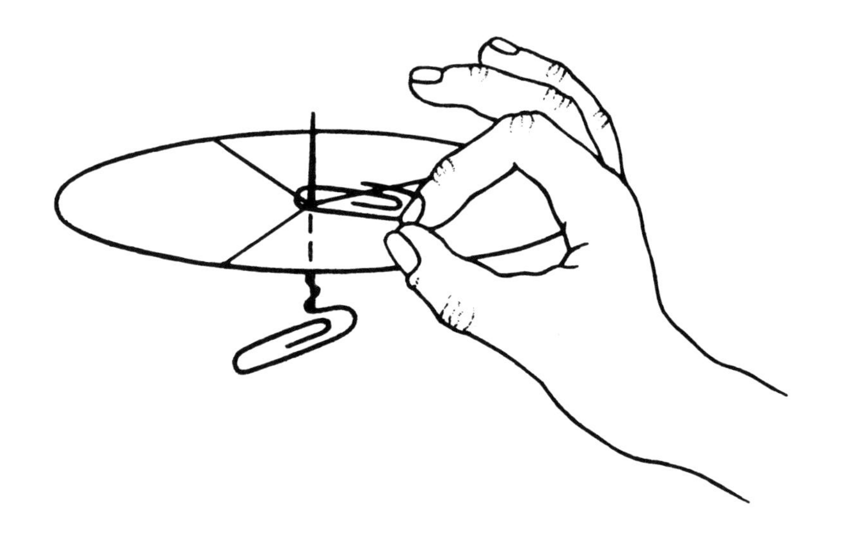

Spinnering

Learning Outcomes

Students will demonstrate an ability to:

- construct a sample space from an event.
- form a simple probability ratio.

Time Requirements

90 minutes

Materials

- Spinnering handouts
- Construction paper
- Two paper clips per team
- Masking tape
- Scissors

NCTM Standards (2000) Addressed

Data Analysis and Probability:

- Select and use appropriate statistical methods to analyze data
- Develop and evaluate inferences and predictions that are based on data
- Understand and apply basic concepts of probability

Prerequisites

Students should be familiar with the concept of sample spaces.

Overview

This group of activities will lead students to understand sample spaces through experimentation. Students will make spinners, then use them to generate sample spaces. The final activity in the group will introduce forming a probability ratio.

Procedures

Part One

Suggested time: 15 minutes

1. Divide class into pairs.
2. Distribute Activity Sheet 1 and other materials. Review and clarify directions. Students prepare spinners according to the directions on the handout.

(continued)

3. As students complete their spinners, check that spinners have been made correctly.

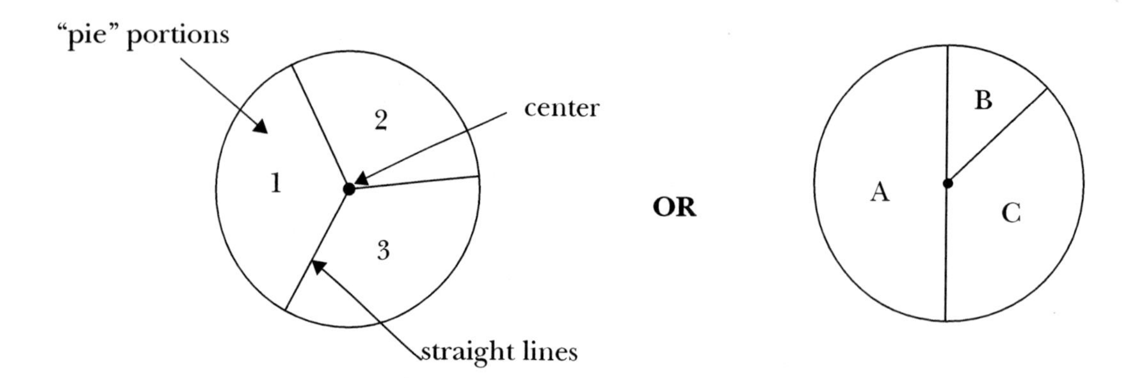

4. Distribute Activity Sheet 2. Go over the directions on the activity sheet, clarifying as needed.

5. Students complete the activity sheet; this should take no more than 15 minutes.

6. As teams complete the activity sheet, assess their work. If some student pairs seem confused, use questioning to focus their responses:

   - "What could the spinners show if they were spun at the same time?"
   - "Have you listed all of the possible combinations?"

Answers to Activity Sheet 2:

First Spinner	Second Spinner
1	A
1	B
1	C
2	A
2	B
2	C
3	A
3	B
3	C

(continued)

Part Two

Suggested time: 10 minutes

7. Ask all students who were directed to use letters for their spinners to meet in one area of the classroom.

   Ask all students who were directed to use numbers for their spinners to meet in another area of the classroom. (If your class has more than twenty students, form four smaller teams by splitting these two large teams in half.)

8. Instruct the teams to share their results and to determine an organized method of displaying their data. Allow about ten minutes for this stage. Identify the "data" as the sample space for this experiment. Use observation and questioning to assess students' understanding. The object is to have students create the sample space display as:

   1X, 2X, . . . , 1Y, 2Y, . . ., etc.

9. After about ten minutes, direct students to return to individual seating.

Part Three

Suggested time: 10 minutes

10. Distribute Activity Sheet 3. Review and clarify the directions. Have students work individually to complete the sample space display. Explain the scoring rubric, making sure students understand how they will be assessed (feedback or grade).

Answers to Activity Sheet 3:

1A 1B 1C 2A 2B 2C 3A 3B 3C

Part Four

Suggested time: 15 minutes

11. Re-form student pairs from the first stage of this activity, then distribute Activity Sheet 4. Review the directions on the handout for completing Column 1; students complete Column 1 as directed.

12. Review and clarify the directions for completing Column 2. As students complete this sheet, use observational assessment to enhance their understanding. Questions for students could include:

(continued)

- "Convince me that this could happen."
- "What were you thinking when you predicted that number?"

Question both accurate and inaccurate results, in order to understand the students' thinking processes.

Part Five

Suggested time: 20 minutes

13. When students have completed Column 2, distribute Activity Sheet 5, the directions for completing Column 3. In this stage of the activity, students work in pairs to spin both spinners 50 times and record the number of times each outcome occurs. Students complete the final part of the activity as directed.

Part Six

Suggested time: 20 minutes

14. Again, have students work in pairs. Distribute Activity Sheet 6, and review and clarify directions. In questions 4 and 7, students are asked to calculate the probability of landing on a consonant as opposed to a vowel. Allow students 15 minutes to complete the activity sheet.

15. Use the scoring rubric at the bottom of Activity Sheet 5. Explain the scoring rubric to students, and be sure they understand how they will be assessed (feedback or grade).

Answers to Activity Sheet 6:

1. $p\,(\text{B}) = \frac{1}{4}$
2. $p\,(\text{C}) = \frac{1}{4}$
3. $p\,(\text{D}) = \frac{1}{4}$
4. $p\,(\text{Consonant}) = \frac{3}{4}$
5. $p\,(\text{B}) = \frac{1}{4}$
6. $p\,(\text{C}) = \frac{1}{4}$
7. $p\,(\text{Consonant}) = \frac{1}{2}$
9. $\frac{1}{6}$

Name________________________________ Date ________________

Spinnering

1. Assign each partner in your team a letter, A or B.

   Letter assigned to you:_______

2. On the construction paper, draw a circle with a radius of 2 inches.

3. With the scissors, cut out the circle.

4. Using the center of the circle as a focus:

   Partner A:

   Draw three regions of any size within your circle. Label the regions 1, 2, and 3. Your spinner may look like this, for example:

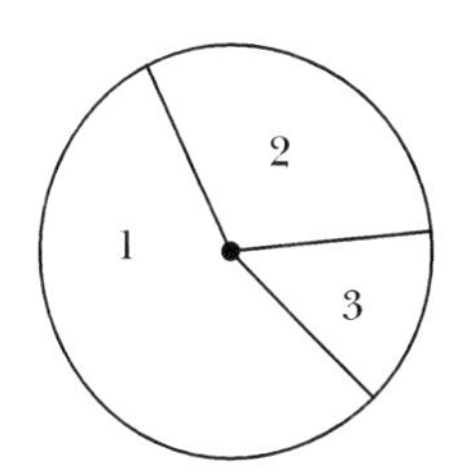

   Partner B:

   Draw three regions of any size. Label the regions A, B, and C. Your spinner may look like this, for example:

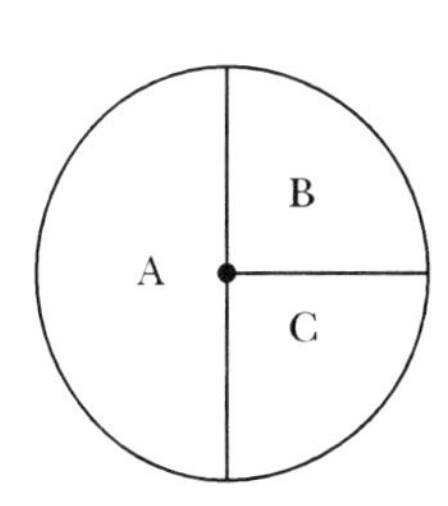

5. Assemble the spinners as shown below.

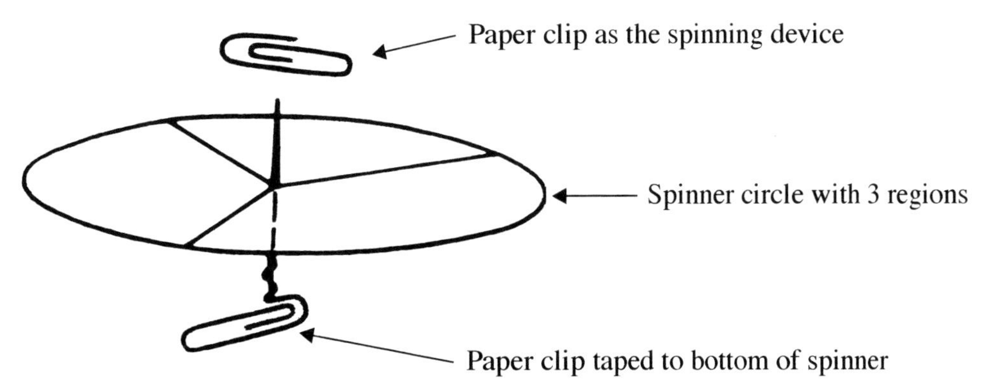

6. Bring the spinners to your teacher to check that they have been made correctly.

Name____________________ Date ____________

ACTIVITY SHEET 2

Spinnering

Draw your spinner in the space below:

Draw your partner's spinner in the space below:

Consider the spinner with the numbers as the first spinner. The spinner with the letters will be the second spinner. In the space below, list all the possible outcomes if both spinners were spun at the same time.

Possible Outcomes

FIRST SPINNER	SECOND SPINNER
____________	____________
____________	____________
____________	____________
____________	____________
____________	____________
____________	____________
____________	____________
____________	____________
____________	____________
____________	____________

Name________________________________ Date ____________________

Spinnering

In the space below, display your **sample space**. Your work will be assessed according to the criteria at the bottom of this page.

PRESENTATION:

0—No work.

1—Work submitted.

2—Work is somewhat neat *or* organized.

3—Work is neat *or* organized.

4—Work is neat *and* organized.

QUALITY (SAMPLE SPACE):

0—No work

1—Work submitted.

2—Some correct outcomes are listed in the sample space.

3—Most outcomes are listed correctly.

4—All outcomes are listed correctly.

Comments: ____

____

____

____

Name____________________ Date ____________

Spinnering

In the first column, list the outcomes from your sample space on Activity Sheet 2.

Suppose you were to spin your spinners 50 times. In how many of those 50 times would you expect each outcome in Column 1 to occur? Write your best estimate in Column 2. Be sure that the numbers you list in Column 2 total 50.

COLUMN 1 OUTCOMES	COLUMN 2	COLUMN 3 (ACTIVITY SHEET 5)
______	______	______
______	______	______
______	______	______
______	______	______
______	______	______
______	______	______
______	______	______
______	______	______
______	______	______

Analysis:

____

____

____

____

____

(continued)

Name________________________________ Date ______________________

Spinnering *(continued)*

1. Spin the spinners 50 times.
2. Record each outcome on the back of this sheet. For example, (1,X) is one of the outcomes.
3. In Column 3 on Activity Sheet 4, record the number of times each outcome actually occurred out of the 50 spins.
4. In the Analysis section on Activity Sheet 4, write down your thoughts about why there were differences between the numbers in Column 2 and the numbers in Column 3.

Your work will be assessed according to the following criteria:

PRESENTATION:

0—No work.

1—Work submitted.

2—Work is somewhat neat *or* somewhat organized.

3—Work is neat *or* organized.

4—Work is neat *and* organized.

QUALITY:

0—No work.

1—Work submitted.

2—Work displays some understanding of the concepts of chance.

3—Work displays an understanding of the concepts of chance, greater chances of occurrence, and the connection between spatial figures and expected outcomes.

4—Work displays achievement beyond expectations.

GRADING TRANSFER

POINT AVERAGE PER OUTCOME:

From (not including) to	Letter grade	From (not including) to	Letter grade
3.1–4.0	A	1.9–2.4	C
3.0–3.1	A–	1.8–1.9	C–
2.9–3.0	B+	1.7–1.8	D+
2.6–2.9	B	1.2–1.7	D
2.5–2.6	B–	1.1–1.2	D–
2.4–2.5	C+	1.1 or below	F

Name____ Date ____________________

Spinnering:
An Introduction to Simple Probability

Probability *(p)* is a measure of how likely an event is. The probability of an event is a number between 0 and 1. Mathematically, it is the ratio of the number of ways an event can happen (*m*) to the number of possible outcomes (*n*). This ratio is written as $p = m/n$.

For example, look at Spinner 1 below. Suppose that you spin the pointer. What is the likelihood that the pointer will stop on A? The probability is $\frac{1}{4}$. The pointer is likely to stop on A one time out of every 4 spins.

Now look at Spinner 2 below. What is the probability that the pointer will stop on A? It is $\frac{1}{2}$. For every 2 spins, it is likely that the pointer will land on A one time.

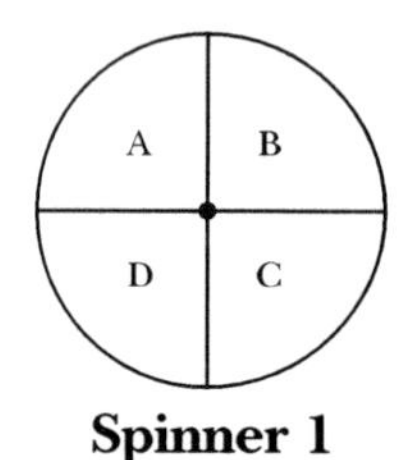

Spinner 1

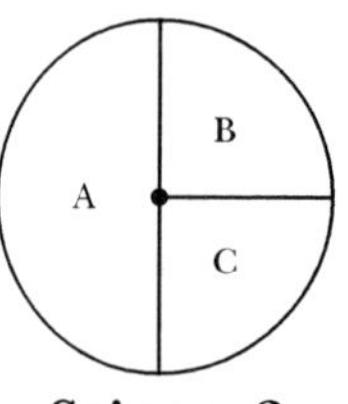

Spinner 2

Find the probabilities *(p)* for each of the following:

Spinner 1

1. *p* (B) = ________________
2. *p* (C) = ________________
3. *p* (D) = ________________
4. *p* (Consonant) = ________________

Spinner 2

5. *p* (B) = ________________
6. *p* (C) = ________________
7. *p* (Consonant) = ________________

8. Draw a spinner with three sections labeled A, B, and C. The outcomes for A, B, and C should produce these probabilities:

   p (A) = $\frac{1}{3}$ *p* (B) = $\frac{1}{2}$

   Use the back of this paper for your drawing.

9. What is the probability of the pointer landing on C?

   p (C) = ____________________________

What's in the Bag?

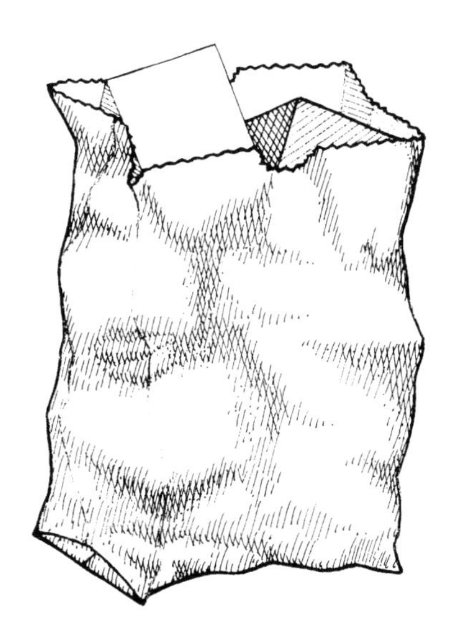

Teacher Page

What's in the Bag?

Learning Outcomes

Students will demonstrate the ability to:

- gather data.
- make predictions and analyses.
- show an exploratory understanding of probability concepts.

Time Requirements

75 minutes

Materials

For each team:

- 24 cubes, in 4 different colors
- A small paper bag
- Reproducible student handouts

NCTM Standards (2000) Addressed

Data Analysis and Probability:

- Develop and evaluate inferences and predictions that are based on data
- Understand and apply basic concepts of probability

Overview

This activity introduces probability concepts through tactile experiences. Student teams receive bags containing colored cubes, then select and replace cubes from the bags, recording and displaying their results. From the data students make predictions about the number of cubes of each color in each bag.

Preparation

Randomly place 24 colored cubes in each bag. Number each bag for easy distribution.

Procedures

Part One

Suggested time: 15 minutes

1. Form an even number of teams, with two or three students in each team. Assign a number to each team.

(continued)

2. Distribute a copy of Tally Sheet 1 to each student. Review and clarify the instructions on the sheet.

3. Students complete the tally sheet as directed.

Part Two

Suggested time: 15 minutes

4. When students have completed Tally Sheet 1, have teams swap bags. Distribute a copy of Tally Sheet 2 to each student. Confirm that the directions are the same as they were for Tally Sheet 1.

5. Students complete the tally sheet as directed.

Part Three

Suggested time: 15 minutes

6. Distribute Reflection Form 1. Review and clarify the directions. Working individually, students use the data they generated in parts 1 and 2 of the activity to complete this sheet. Remind them that they will be assessed according to their ability to gather and display data and to make predictions, following the rubric on the form.

Part Four

Suggested time: 15 minutes

7. Distribute the original bags to the original teams, and distribute Tally Sheet 3. Review and clarify the instructions. Students work in teams to complete the sheet.

Part Five

Suggested time: 15 minutes

8. Distribute Reflection Form 2. Review and clarify the directions. Have students work individually to complete the sheet. Tell students that their work will be assessed using the same scoring rubric found on Reflection Form 1.

Name____ Date ____________________

TALLY SHEET 1

What's in the Bag?

Team Number ________ Bag Number ________

Keep the bag closed so that no one can see inside. Each team member will keep his or her own data chart.

1. Randomly select one cube at a time from the bag. (NO PEEKING!) Write the color of the cube below. Replace the cube in the bag. Do this ten times.

   ________ ________ ________ ________ ________

   ________ ________ ________ ________ ________

2. There are 24 cubes in the bag. Without talking with your teammate(s), write your predictions for the number of cubes of each color you think are in the bag. Use the lines below.

COLOR	NUMBER IN THE BAG
____________	____________
____________	____________
____________	____________
____________	____________

3. Share with your teammate(s) the predictions you made above. As a team, make predictions all members can agree on. Write the team predictions below.

COLOR	NUMBER IN THE BAG
____________	____________
____________	____________
____________	____________
____________	____________

Name________________________________ Date ______________

TALLY SHEET 2

What's in the Bag?

Team Number ________ Bag Number ________

Keep the bag closed so that no one can see inside. Each team member will keep his or her own data chart.

1. Randomly select one cube at a time from the bag. (NO PEEKING!) Write the color of the cube below. Replace the cube in the bag. Do this ten times.

   ________ ________ ________ ________ ________

   ________ ________ ________ ________ ________

2. There are 24 cubes in the bag. Without talking with your teammate(s), write your predictions for the number of cubes of each color you think are in the bag. Use the lines below.

COLOR	NUMBER IN THE BAG
________	________
________	________
________	________
________	________

3. Share with your teammate(s) the predictions you made above. As a team, make predictions all members can agree on. Write the team predictions below.

COLOR	NUMBER IN THE BAG
________	________
________	________
________	________
________	________

Name________________________________ Date ________________

What's in the Bag?

Think about the data you collected on both tally sheets. What predictions would you make about the number of cubes of each color in the first bag your team received? Explain your data tables, and your predictions, on the back of this sheet. Your work will be assessed using the scoring rubric below.

SCORING RUBRIC

0—No work.
1—Work submitted.
2—Work shows some achievement of expected outcomes.
3—Work shows achievement of all expected outcomes.
4—Work displays achievement beyond expectations.

DESCRIPTORS

Data gathering: The data are accurate and displayed clearly and in a mathematical manner.

Predictions: Evidence is presented showing that the predictions were based on data and were supported with accurate statements.

SCORES

Data Gathering = ________
Predictions = ________
Average Score = ________

GRADING TRANSFER

POINT AVERAGE PER OUTCOME:

From (not including) to	Letter grade	From (not including) to	Letter grade
3.1–4.0	A	1.9–2.4	C
3.0–3.1	A–	1.8–1.9	C–
2.9–3.0	B+	1.7–1.8	D+
2.6–2.9	B	1.2–1.7	D
2.5–2.6	B–	1.1–1.2	D–
2.4–2.5	C+	1.1 or below	F

Name________________________________ Date ________________

What's in the Bag?

Combine your data with the data of the team with whom you exchanged bags.

1. Consider only the data from the first bag assigned to *your* team. By yourself, predict the number of cubes of each color in the first bag. Record your predictions below.

COLOR	NUMBER IN THE BAG
________________	________________
________________	________________
________________	________________
________________	________________

2. By yourself, predict the number of cubes of each color in the bag first assigned to your "partner" team.

COLOR	NUMBER IN THE BAG
________________	________________
________________	________________
________________	________________
________________	________________

Name____ Date ____________________

What's in the Bag?

What is your prediction about the numbers of cubes of each color in each of the bags about which you gathered data? Use the assessment feedback from this series of activities and the results from Tally Sheet 3.

BAG #	COLOR/NUMBER OF CUBES
________	________ ________ ________ ________ ________
________	________ ________ ________ ________ ________

Write an argument in support of your prediction. Use the space below.

____

____

____

____

____

____

____

____

____

____

____

____

____

The Union of Intersections

The Union of Intersections

Teacher Page

Learning Outcomes

Students will display the ability to:

- compute the probability of an event composed of non-exclusive subevents.
- gather data.
- record data.
- calculate simple probability.

Time Requirements

1 hour 50 minutes

Materials

Reproducible student handouts

NCTM Standards (2000) Addressed

Data Analysis and Probability:

- Develop and evaluate inferences and predictions that are based on data
- Understand and apply basic concepts of probability

Prerequisites

Students need to know how to compute simple probabilities.

Overview

This series of activities is an opportunity to explore and identify concepts. After gathering data, student teams will determine the probabilities of simple events based on more than one data set. The key to finding a correct probability lies in students realizing that an outcome appearing in two unique sets cannot be counted twice in the computation of a probability. Assessment through observation will be key to understanding which teams (students) have realized this and which have not.

Procedures

In Part 1 of this activity, student teams will gather data from classmates, record data on a tally sheet, calculate simple probabilities, and share the results with the entire class. In Part 2, students will compute probabilities for "and" and "or" prompts.

(continued)

Part One

Suggested time: 1 hour 20 minutes

1. Divide the class into six teams, each with about the same number of students. Assign a number from 1 to 6 to each team. Distribute the separated portions of Activity Sheet 1 to each team. Go over the directions with the class.

2. Distribute Activity Sheet 2, the Tally Sheet. Be sure all student teams know how to record their data on this sheet.

3. If you wish, share the prompts for all teams with the class. This simplifies the data-gathering process, as all students know in advance the information required.

4. Distribute Activity Sheet 3. This portion of the activity should be completed in teams.

5. Collect the completed Activity Sheet 3 to assess students' understanding of simple probability. The activity sheet contains some simple probability prompts, some items containing "or" prompts, and some items containing "and" prompts. A review of the completed activity sheets will help you determine whether students are able to compute probability and have an emerging understanding of the computation of "and" and "or" related probability prompts.

6. Give each student a copy of Activity Sheet 4, the Class Tally Sheet. Choose one student from each team to present the team's data. All students should record the data for each prompt.

Part Two

Suggested time: 30 minutes

7. Divide the class into teams of four. As far as possible, mix students from the teams in Part 1, so that each new team includes students who worked on different prompts.

8. Distribute Activity Sheet 5. Explain and clarify the directions. Students proceed as directed on the handout. Tell students that they will be assessed according to the criteria listed at the bottom of the activity sheet.

The Union of Intersections

TEAM NUMBER 1

Each team will gather data from students in the class. Each data set will be different. Teams will share their collected data with the entire class. Use Activity Sheet 2 to record your team's data.

Your team's **prompt:**

Determine which students in the class are wearing clothes of the following colors:

White—use letter "A" on the tally sheet
Red —use letter "B" on the tally sheet
Blue —use letter "C" on the tally sheet

- - -

The Union of Intersections

TEAM NUMBER 2

Each team will gather data from students in the class. Each data set will be different. Teams will share their collected data with the entire class. Use Activity Sheet 2 to record your team's data.

Your team's **prompt:**

Determine which students are wearing the clothing listed below:

Jeans —use letter "A" on the tally sheet
Sweatshirt—use letter "B" on the tally sheet
T-shirt —use letter "C" on the tally sheet

- - -

The Union of Intersections

TEAM NUMBER 3

Each team will gather data from students in the class. Each data set will be different. Teams will share their collected data with the entire class. Use Activity Sheet 2 to record your team's data.

Your team's **prompt:**

Determine which students in the class have the following physical characteristics:

Left-handed—use letter "A" on the tally sheet
Black hair —use letter "B" on the tally sheet
Brown eyes —use letter "C" on the tally sheet

- - -

(continued)

The Union of Intersections

TEAM NUMBER 4

Each team will gather data from students in the class. Each data set will be different. Teams will share their collected data with the entire class. Use Activity Sheet 2 to record your team's data.

Your team's **prompt:**

Determine which students have the following:

A winter birthday —use letter "A" on the tally sheet
More than one sister —use letter "B" on the tally sheet
At least one younger sibling—use letter "C" on the tally sheet

The Union of Intersections

TEAM NUMBER 5

Each team will gather data from students in the class. Each data set will be different. Teams will share their collected data with the entire class. Use Activity Sheet 2 to record your team's data.

Your team's **prompt:**

Determine which students in the class have more than six letters in their:

First name —use letter "A" on the tally sheet
Middle name—use letter "B" on the tally sheet
Last name —use letter "C" on the tally sheet

The Union of Intersections

TEAM NUMBER 6

Each team will gather data from students in the class. Each data set will be different. Teams will share their collected data with the entire class. Use Activity Sheet 2 to record your team's data.

Your team's **prompt:**

Determine which students in the class have telephone numbers that contain:

At least one "3"—use letter "A" on the tally sheet
At least one "4"—use letter "B" on the tally sheet
At least one "5"—use letter "C" on the tally sheet

Name________________________________ Date ________________

Activity Sheet 2
Tally Sheet

Team Number ______

The Union of Intersections

On the chart below, write the names of classmates who responded "yes" to the prompts on your team's assignment.

A	B	C

Name____________________ Date ____________

TEAM NUMBER ______

The Union of Intersections

Work with your teammates to determine the probabilities below. Use the data you collected.

1. p (A) = ______________

2. p (B) = ______________

3. p (C) = ______________

4. Write in words: p(A or B)

   ____

5. Write in words: p(A and B)

   ____

6. Write in words: p(A, B, or C)

   ____

7. Write in words: p(A, B, and C)

   ____

Calculate:

8. p (A or B) = ______________

9. p (A and B) = ______________

10. p (A, B, or C) = ______________

11. p (A, B, and C) = ______________

12. On the back of this sheet, explain the difference between "and" and "or" probabilities.

Name________________________________ Date ______________

The Union of Intersections

Use this form to record the data for each team's prompt.

TEAM 1 PROMPT—
DETERMINE WHICH STUDENTS IN THE CLASS ARE WEARING CLOTHES OF THE FOLLOWING COLORS:

WHITE—A	RED—B	BLUE—C

TEAM 2 PROMPT—
DETERMINE WHICH STUDENTS ARE WEARING THE ITEMS OF CLOTHING LISTED BELOW:

JEANS—A	SWEATSHIRT—B	T-SHIRT—C

(continued)

Name_______________________________ Date ________________

ACTIVITY SHEET 4
CLASS TALLY SHEET

The Union of Intersections *(continued)*

Use this form to record the data for each team's prompt.

TEAM 3 PROMPT—
DETERMINE WHICH STUDENTS IN THE CLASS HAVE THE FOLLOWING PHYSICAL CHARACTERISTICS:

LEFT-HANDED—A	BLACK HAIR—B	BROWN EYES—C

TEAM 4 PROMPT—
DETERMINE WHICH STUDENTS HAVE THE FOLLOWING:

A WINTER BIRTHDAY—A	MORE THAN ONE SISTER—B	AT LEAST ONE YOUNGER SIBLING—C

(continued)

Name________________________ Date ____________

ACTIVITY SHEET 4
CLASS TALLY SHEET

The Union of Intersections *(continued)*

Use this form to record the data for each team's prompt.

TEAM 5 PROMPT—
DETERMINE WHICH STUDENTS IN THE CLASS HAVE MORE THAN SIX LETTERS IN THEIR:

FIRST NAME—A	MIDDLE NAME—B	LAST NAME—C

TEAM 6 PROMPT—
DETERMINE WHICH STUDENTS IN THE CLASS HAVE TELEPHONE NUMBERS THAT CONTAIN:

AT LEAST ONE "3"—A	AT LEAST ONE "4"—B	AT LEAST ONE "5"—C

Name____ Date ____________________

The Union of Intersections

Work with your teammates to determine the probabilities listed on this page and the next. Use the data collected about this class. Each teammate should complete an individual activity sheet. Show all work! You will be assessed and graded according to the criteria listed at the bottom of the next page.

1. p(white clothes) = __________
   WORK:

2. p(black hair or blue clothes) = __________
   WORK:

3. p(Having a first name with at least six letters or a last name with at least six letters) = __________
   WORK:

4. p(Being left-handed and having a telephone number containing at least one "5") = __________
   WORK:

(continued)

Name________________________________ Date ______________

The Union of Intersections *(continued)*

5. p(Having at least one younger sibling, wearing a T-shirt, or having a telephone number with a "5" in it) = __________

WORK:

6. p(Having a last name with more than 6 letters, a winter birthday, and brown eyes) = __________

WORK:

ANSWERS

0—None shown.
1—Answers displayed.
2—Responses include some correct answers.
3—Most answers are correct.
4—All answers are correct.

WORK

0—None shown.
1—Some responses contain student work.
2—All work is displayed.
3—Work demonstrates expected understanding of the skill of computing probability.
4—Work demonstrates understanding beyond expectations.

PRESENTATION

0—No work is displayed.
1—Work is displayed.
2—Work is somewhat neat *or* organized.
3—Work is neat *or* organized.
4—Work is neat *and* organized.

SCORES

Presentation Score = ______ × 2 = ________
Work Score = ______ × 4 = ________
Answer Score = ______ × 6 = ________
Total (____) + 52 = ______ (grade)

Comments: ____

Compound Events

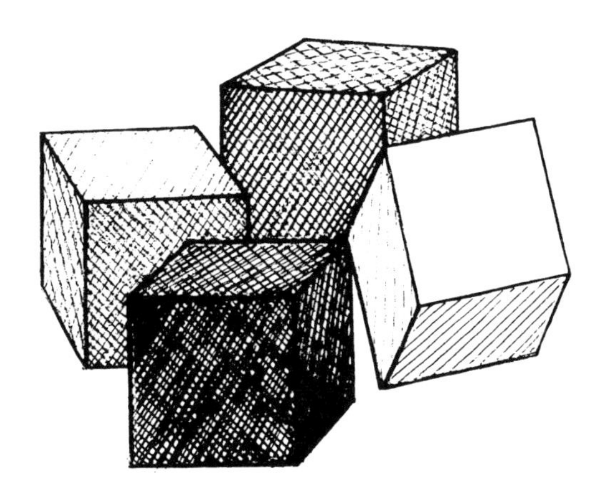

Compound Events

Teacher Page

Learning Outcomes

Students will demonstrate the ability to:

- compute the probability of compound events.
- gather data.
- make predictions and analyses.

Time Requirements

2 hours 10 minutes

Materials

- Reproducible student handouts
- Small paper bags (one for every team)
- Red, blue, and white cubes (24 cubes per team)
- Construction paper (one sheet for each spinner)
- Paper clips (two per spinner)
- Masking tape
- Scissors

NCTM Standards (2000) Addressed

Data Analysis and Probability:

- Develop and evaluate inferences and predictions that are based on data
- Understand and apply basic concepts of probability

Prerequisites

Students should know how to construct a sample space and compute simple probabilities.

Overview

This series of activities represents the exploration and concept-identification levels of learning. The activity sheets are designed to lead students to the discovery that $p(\text{A}) \times p(\text{B}) = p(\text{A followed by B})$. The first and second parts of the activity involve the use of spinners. The third activity involves the random selection of colored cubes from a paper bag, with a focus on "with replacement" and "without replacement."

Preparation

For Part 3 of the activity, randomly place 24 cubes in each paper bag: some red, some white, and the rest blue. Write a number on each bag. On a separate sheet, note the number of cubes of each color placed in each bag so that the exact probabilities can be assessed for each team.

(continued)

Procedures

Part One

Suggested time: 30 minutes

1. Form teams of three to four students. Distribute Activity Sheet 1. Each student should complete his or her own copy of the activity sheet.
2. When all teams have completed Activity Sheet 1, ask one student from each team to share that team's responses with the rest of the class.
3. After all teams have presented their responses, have students return to their teams and edit their responses if they think it appropriate.
4. Collect student work. Assess to see if students generated the correct responses. If the responses were generally correct, continue with Part 2 of the activity. If the responses were generally incorrect, lead a class discussion to synthesize a correct analysis.

Answers to Activity Sheet 1:

1. (a) $\frac{1}{2}$ (b) $\frac{1}{4}$ (c) $\frac{1}{4}$
2. AA, AB, AC, BA, BB, BC, CA, CB, CC
3. (a) $\frac{1}{8}$ (d) $\frac{1}{8}$ (g) $\frac{1}{8}$
   (b) $\frac{1}{4}$ (e) $\frac{1}{16}$ (h) $\frac{1}{16}$
   (c) $\frac{1}{16}$ (f) $\frac{1}{8}$ (i) $\frac{1}{16}$

Part Two

Suggested time: 30 minutes

5. Form new teams of three (four if necessary, but as few teams of four as possible). Distribute Activity Sheet 2 and the materials for making spinners to each team (one spinner per team member).
6. Assign each member of the team a letter, A, B, or C (for teams for four, repeat one of the letter assignments). Students prepare spinners as directed on the handout and then answer the questions.

(continued)

Part Three

Suggested time: 40 minutes

7. Randomly form new teams of three or four students. Distribute copies of Activity Sheet 3 and a bag of cubes to each team. Note the number of the bag assigned to each team. Review and clarify the directions. Students proceed as directed on the handout.

8. Collect and assess student work. Return the work to the students. In a class discussion, review the points of "with replacement" and "without replacement."

Part Four

Suggested time: 30 minutes

The last part of this series is a culminating assessment opportunity. The problem presented here requires high-level thinking. Most students will get a good start if allowed to converse with other students before writing their own solutions.

9. Form random teams of four. Distribute a copy of the Assessment Activity to each student. Allow students ten minutes to brainstorm how they would solve the problem. Encourage students to take notes to help them formulate a response. After the ten minutes elapse, students complete the assessment activity individually. Tell them that they will be evaluated according to the Assessment Criteria sheet that follows the Assessment Activity. Be sure to clarify all criteria with students before proceeding.

Name____ Date ____________________

Compound Events

With your teammates, complete the following:

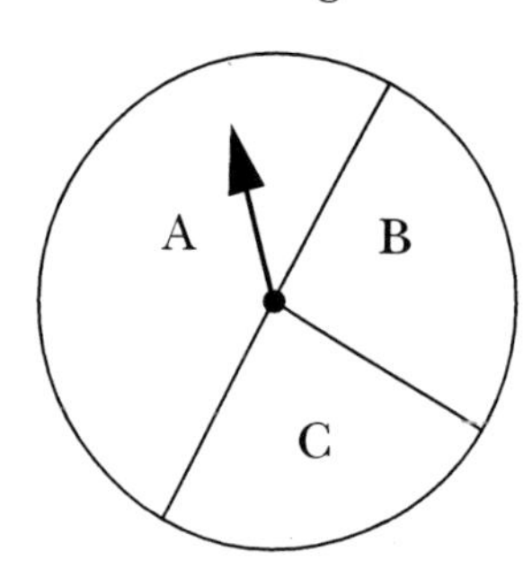

1. If this spinner were spun once, what would the probability be for each of the following?

   (a) p(A)= ______ (b) p(B)= ______ (c) p(C)= ______

2. If the spinner were spun twice and the results of each pair of spins was recorded, what would all the possible outcomes be? List the outcomes in the sample space.

   ____

   ____

3. Brainstorm with your teammates to find the probabilities of these events.

   (a) p(A then B) = ________ (f) p(A then C) = ________

   (b) p(A then A) = ________ (g) p(B then A) = ________

   (c) p(B then C) = ________ (h) p(B then B) = ________

   (d) p(C then A) = ________ (i) p(C then C) = ________

   (e) p(C then B) = ________

4. Describe a mathematical method to compute the correct response to the examples presented in questions 3. Use the back of this activity sheet for your answer.

Name______________________________ Date ______________

Compound Events

Letter assigned to you: ____

1. Using the materials, make a spinner as directed below.

   Team member A—Make a spinner on which the outcomes are A, B, and C.

   Team member B—Make a spinner on which the outcomes are 1, 2, and 3.

   Team member C—Make a spinner on which the outcomes are Q and R.

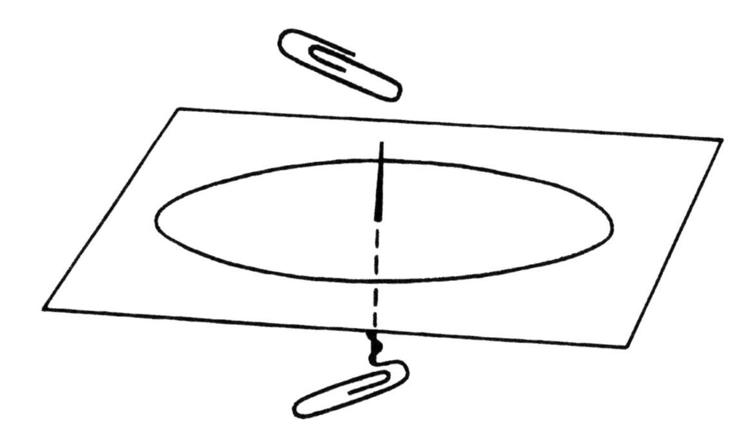

2. Determine the probability for each outcome on the spinner you made. List them below.

   ____

3. Suppose all three spinners were spun at the same time. What are all the possible outcomes? List them below.

   ____

   ____

   ____

   Next to each outcome write the probability for its occurrence. If you need more room, use the back of this sheet.

Name________________________________ Date ________________

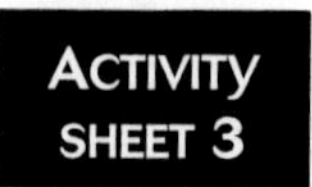

Compound Events

Bag Number _____

With your teammates, determine the answers for each of the following:

1. Inspect the contents of the bag assigned to your team. Determine the probabilities for each of the following:

p(red cube) = ________ p(blue cube) = ________ p(white cube) = ________

2. Randomly select two cubes from the bag. Select the first cube and place it on the table. Then select a second cube and place it beside the first cube. In the space provided, write the sample space.

3. Determine the probabilities for each of the following:

(a) p(red, first) = ______ p(white, second) = ______ p(red then white) = ______

(b) p(blue, first) = ______ p(red, second) = ______ p(blue then red) = ______

(c) p(white, first) = ______ p(blue, second) = ______ p(white then blue) = ______

4. Consider a random selection of cubes using the following rules: The first cube is selected and the color is recorded. The cube is replaced in the bag. A second cube is selected and the color is recorded. Would this change in the procedure change the probabilities computed in # 3 above? In the space below, explain your answer in detail.

____

____

____

____

Name____ Date ____________________

Compound Events

You have a friend who has never worked with probability of compound events. In a letter to you she presents this problem:

> There are 10 green marbles, 5 black marbles, and 15 silver marbles in a bag. Two marbles are selected, one at a time and randomly. The first marble is not returned to the bag before the second marble is selected. Is it more likely that two marbles of the same color will be selected? Or two marbles of different colors?

Write back to your friend. Explain to her how to do this problem. Continue your answer on the other side of this sheet if you need more space.

____

____

____

____

____

____

____

____

____

____

____

____

____

____

____

Name________________________________ Date ______________

Compound Events

Your work will be assessed according to the criteria:

1. The work is neat and organized.
2. The work demonstrates an expected achievement of the following learning outcomes:
   - correctly computed simple probabilities
   - a correctly displayed sample space
   - correctly calculated probabilities for compound events

CONTENT

0—No work submitted.
1—Work submitted.
2—Work demonstrates some achievement of outcome.
3—Work demonstrates expected achievement of learning outcome.
4—Work demonstrates achievement beyond expectations.

PRESENTATION

0—No work submitted.
1—Work submitted.
2—Work is somewhat neat *or* organized.
3—Work is neat *or* organized.
4—Work is neat *and* organized.

SCORES

Presentation = ______
Content outcome 1 = ______
Content outcome 2 = ______
Content outcome 3 = ______
Average score (nearest 10th) = ______

GRADING TRANSFER

POINT AVERAGE PER OUTCOME:

From (not including) to	Letter grade	From (not including) to	Letter grade
3.1–4.0	A	1.9–2.4	C
3.0–3.1	A–	1.8–1.9	C–
2.9–3.0	B+	1.7–1.8	D+
2.6–2.9	B	1.2–1.7	D
2.5–2.6	B–	1.1–1.2	D–
2.4–2.5	C+	1.1 or below	F

Comments ____

____

____

M Cubed

M Cubed

Teacher Page

Learning Outcome

Students will demonstrate an ability to:

- generate data sets given measures of central tendency.

Time Requirements

1 hour 40 minutes

Materials

Reproducible student handouts

NCTM Standards (2000) Addressed

Data Analysis and Probability:

- Formulate questions that can be addressed with data, and collect, organize, and display relevant data to answer them
- Select and use appropriate statistical methods to analyze data

Prerequisites

Students should have some knowledge of mean, median, and mode.

Overview

This activity series assesses students' understanding of mean, median, and mode. Given these measures of central tendency, teams of students will be challenged to make an accurate data set. Given data sets, students will be challenged to determine the measures of central tendency and to identify which of those measures best represent the data set. Explanations of student choices will be required.

Procedures

Part One

Suggested time: 30 minutes

1. Briefly review the definitions of *mean, median,* and *mode* with students. Make sure that they are all comfortable with these concepts before proceeding to the first activity.

(continued)

2. Randomly form teams of three to four students. Distribute Activity Sheet 1. Review and clarify the directions. Allow 10 minutes for brainstorming.
3. Reassign students to individual seating, and direct students to complete their responses to the prompts on Activity Sheet 1. This should take about 20 minutes.
4. Collect, assess, and return student work.

Part Two

Suggested time: 30 minutes

5. Randomly form teams of three to four students. Distribute Activity Sheet 2. Explain and clarify the directions. Again, allow 10 minutes for brainstorming, then direct students to complete the activity sheet individually. Given the data set, students should describe and explain which of the measures of central tendency best describe the data.

Answers to Activity Sheet 2:

1. Mean = 5
   Median = 2
   Mode = 1
   The median should be circled.
2. Mean = $80\frac{1}{3}$
   Median = 85
   Mode = None
   The median should be circled.

Part Three

Suggested time: 40 minutes

6. Form random teams of three to four students. Distribute Activity Sheet 3. Review the instructions on the activity sheet. Again, allow 10 minutes for brainstorming, then direct students to complete the activity sheet individually. Tell them that they will be evaluated according to the Assessment Criteria Sheet that follows Activity Sheet 3. Be sure to clarify all criteria with students before proceeding.

Name______________________________ Date ____________________

M Cubed

First, brainstorm with your teammates about how to respond to the prompts below. Then, working on your own, formulate your responses. Use the space on the back of this sheet for your answers. You will be assessed according to the criteria at the bottom of this page.

1. Form a set of data that meets the following criteria:
   - The data set has seven numbers.
   - The mode is 1.
   - The median is 3.
   - The mean is 4.

2. Form a set of data that meets the following criteria:
   - The set has 10 numbers.
   - The median is 6.
   - The mean is 6.
   - The number 6 is not in the data set.
   - All numbers in the data set are the modes.

PRESENTATION

0—No work submitted.
1—Work submitted.
2—Work is somewhat neat *or* organized.
3—Work is neat *or* organized.
4—Work is neat *and* organized.

CONTENT

0—No evidence of work submitted.
1—Work submitted.
2—Some evidence of understanding present.
3—Expected understanding of how to calculate the measure of central tendency is presented.
4—Work includes a complete explanation of the process by which the measure of central tendency is determined.

ASSESSMENT RATING

Topic	Score
Mean	_________
Median	_________
Mode	_________

Topic	Score
Presentation	_________
Average Score	_________

GRADING TRANSFER

AVERAGE SCORE:

From (not including) to	Letter grade	From (not including) to	Letter grade
3.1–4.0	A	1.9–2.4	C
3.0–3.1	A–	1.8–1.9	C–
2.9–3.0	B+	1.7–1.8	D+
2.6–2.9	B	1.2–1.7	D–
2.5–2.6	B–	1.1–1.2	E
2.4–2.5	C+	1.1 or below	F

Name________________________________ Date ______________

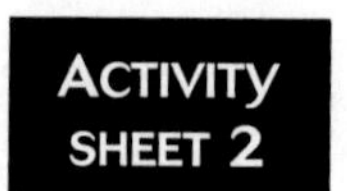

M Cubed

1. From the following data set, identify the mean, median, and mode.

   {1, 2, 9, 7, 1}

   Mean = _________

   Median = _________

   Mode = _________

   Circle the statistic—mean, median or mode—that best represents the data.

2. From the following data set—identify the mean, median, and mode.

   {98, 79, 80, 90, 95, 40}

   Mean = _________

   Median = _________

   Mode = _________

   Circle the statistic—mean, median, or mode—that best represents the data.

3. Write an explanation for both choices. Use the space below or the back of this sheet.

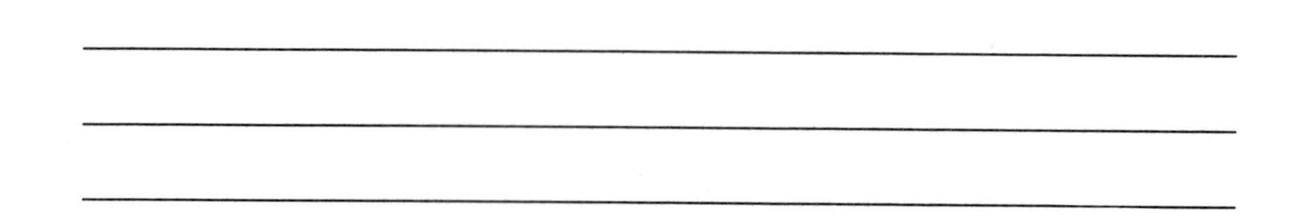

Name________________________________ Date ________________

M Cubed

1. Make a data set that meets the following criteria:
   - The set contains 5 numbers.
   - The **mean** is the measure of central tendency that best reflects the data.

   Data Set = {   }

2. Make a data set that meets the following criteria:
   - The set contains 5 numbers.
   - The **median** is the measure of central tendency that best reflects the data.

   Data Set = {   }

3. Make a data set that meets the following criteria:
   - The set contains 5 numbers.
   - The **mode** is the measure of central tendency that best reflects the data.

   Data Set = {   }

4. In the chart below, explain your responses to each of the prompts above.

DATA SET	MEASURE OF CENTRAL TENDENCY	EXPLANATION
1	Mean	
2	Median	
3	Mode	

Name____________________ Date ____________

M Cubed

Assessment for this activity will be determined by the following rubrics and grade transfer guidelines:

PRESENTATION

0—No work submitted.
1—Work submitted.
2—Work demonstrates some achievement of the expected outcomes.
3—Work demonstrates some achievement of the expected outcomes.
4—Work demonstrates achievement beyond expected outcomes.

OUTCOMES

- **Presentation** (neatness and organization)
- Correct **data** for each activity
- Appropriate **reasoning** for the rationale used in responses

SCORES

Presentation = ________
Reasoning = ________
Data = ________
Average = ________

GRADING TRANSFER

POINT AVERAGE PER OUTCOME:

From (not including) to	Letter grade	From (not including) to	Letter grade
3.1–4.0	A	1.9–2.4	C
3.0–3.1	A–	1.8–1.9	C–
2.9–3.0	B+	1.7–1.8	D+
2.6–2.9	B	1.2–1.7	D
2.5–2.6	B–	1.1–1.2	D–
2.4–2.5	C+	1.1 or below	F

Surveying and Sampling

Surveying and Sampling

Learning Outcomes

Students will demonstrate an understanding of:

- creating a survey.
- sampling methods.

Time Requirements

5 hours

Materials

- Reproducible student handouts
- Chart paper

NCTM Standards (2000) Addressed

Data Analysis and Probability:

- Formulate questions that can be addressed with data, and collect, organize, and display relevant data to answer them
- Select and use appropriate statistical methods to analyze data

Overview

This series of activities has students identify a topic of inquiry, generate topic questions to be pursued, create survey items from this topic focus, execute a survey, and analyze data.

Preparation

You will need first to obtain approval for surveying the school's student population. Then establish a time and place for carrying out the survey. Coordination with English/ Language Arts teachers would facilitate integrating subject areas within this activity and enhance the student product.

Procedures

Part One

Suggested time: 45–60 minutes

1. In a class discussion, inform the class that they are about to conduct a survey of their classmates throughout the school. Ask students about their understanding of

(continued)

surveying. Discuss the connection between the results of surveying and decision-making processes in the public and private sector. A discussion about political polls, television surveys, telemarketing, and other current trends will help students understand why they are pursuing this topic.

2. Randomly form teams of three or four students. Distribute Activity Sheet 1. Explain and clarify the directions.

3. After students have completed Activity Sheet 1, collect their responses. Identify those topics that are appropriate for a school-based investigation. Select a topic in which the entire class has some interest.

4. Distribute Activity Sheet 2. Explain and clarify the directions. This activity sheet is designed to help students focus the survey and identify the types of questions appropriate for a survey.

5. Use the reverse jigsaw method to have students share what each team has written. Assign a letter to each member of each team: A, B, C, or D. Instruct all students assigned the same letter to meet in a common area and discuss what they have written. After 15 minutes, direct the students to return to their "home" team and to share what they have heard and edit their current responses (if they so desire).

6. Collect and review the responses on Activity Sheet 2. Lead a class discussion that focuses on the items that the students feel are the most important and appropriate for a school-based activity. Limit the number of items for the survey to 10 or fewer.

Part Two

Suggested time: 2 hours

In this part of the activity, students will write the survey items. To enhance the effectiveness of this activity, model various types of survey items. For example:

- **To show choice:**

  The amount of hours I listen to rap music each week can best be described as: *(circle one)*

  0 hours 2 hours 4 hours 6 hours 8 hours more than 8 hours

- **To show preference:**

  I think rap music is great. *(circle one)*

  Strongly disagree Disagree Mildly disagree Agree Strongly agree

- **To show background of those responding:**

  Most of the grades I receive in school are: *(circle one)*

  A B C D F

(continued)

7. Form new teams, one for each of the survey items. Assign one of the survey items to each team.

8. Distribute Activity Sheet 3. Review and clarify the directions. Collect and assess Activity Sheet 3. Assessment for this activity sheet is informal, but should provide valuable information about how the students are progressing toward writing survey items. Write comments in a narrative form on the students' sheets. Assessment by an English/Language Arts teacher would provide more feedback to the students and would assist them in formatting well-written survey items.

9. Return papers and re-form teams from Step 7. Allow time for teams to review comments from the assessment feedback. Encourage students to edit their work accordingly.

10. Select one student from each team to write the team's "final" version of the item assigned on the chalkboard (or other central location). After all of the items have been offered, critique them with the class. Include questions like:

    - Are these items going to generate the responses the students are seeking?

    - Are the items worded so that the person completing the survey can understand it?

    - Are the response choices appropriate and complete?

11. Discuss the potential for correlating the item responses as a part of the analysis. (Using the example in the modeling suggestion, students might try to correlate the number of hours rap music is listened to and whether or not the person responding likes rap music.)

12. After class discussion has been completed and students agree on the wording and content of the items offered by each student team and the items which may lend themselves to an analytic correlation, copy the survey and distribute it to each student.

Part Three

Suggested time: 1 hour

This part of the activity will introduce students to methods of surveying. Random sampling is the focus of the learning. Each team will be given a sampling method. The students will critique and discuss their view of its validity. One sampling method will be agreed upon by the class. The focus is that the sample be as "random" as possible. It is important to lead the students in that direction as the activity series continues.

(continued)

13. Randomly form six new teams of students. Copy and cut Activity Sheet 4 along the dotted lines. Assign each team a number. Distribute the activity sheet sections to the correspondingly numbered teams. Distribute a piece of chart paper to each team. Have students proceed according to the directions on their team's sheets. This portion should take about 30 minutes.

14. Direct each team to transcribe its results to the chart paper. Post the chart paper on a wall. Each team should review all of the other teams' responses. In a class discussion, come to a consensus about the appropriateness of each item. Focus on the desirableness of a random sample.

    Method 1: Somewhat limited. It may be exclusive to a nationality.

    Method 2: Possibly good, but may not generate a random sample. The first students may all ride the same bus.

    Method 3: May not generate a random sample. What if a specific population is in study hall or other assignment during second period?

    Method 4: Possibly good. What if some students eat outside the cafeteria? Do they represent a specific population which may not be surveyed using this method?

    Method 5: Excellent method. Logistics need to be developed.

    Method 6: Potentially biased, ethnically and racially.

15. Once the class agrees on a credible sampling method, and after receiving appropriate approval from the school administration, conduct the survey.

    Things to double-check before proceeding:

    - Make sure enough copies of the survey are available.
    - Make sure every student has a surveying assignment that does not conflict with his or her schedule.
    - Make sure that all students have a similar understanding of their mission.

Part Four

Suggested time: 30 minutes

16. Gather all the completed surveys and conduct class meetings to facilitate the documentation (e.g., graphing) of the data. With the knowledge of the survey prompts, determine the next foci appropriate for this activity series. These may include:

(continued)

- Is there a correlation possible among responses to survey items that may have significant meaning?
- What graphical display(s) would be most appropriate to display the data and to support the analysis of data?
- What errors in the data gathering, survey instrument, and analysis could have been made?
- What interpretations can be made from the gathered data?

Part Five

Suggested time: 30 minutes

17. Distribute Activity Sheet 5, which provides a self-assessment prompt for this series of activities. Students should complete this sheet individually. Scoring and grading of this activity sheet can be based on the criteria listed on the page following Activity Sheet 5.

Name____________________ Date ____________

Surveying and Sampling

With your teammates, discuss topics that you think:

- are important to your school
- would be better understood by asking questions of the student body
- could provide some ideas about the future direction of your school

What are some topics that you think your school could benefit by knowing more about? List them below.

Name____ Date ____________________

Surveying and Sampling

Write the topic for the class survey on the line below.
____

Think of five different items about this topic that you can learn about through a survey. Work with your teammates. List the items below.

1. ____

____

____

2. ____

____

____

3. ____

____

____

4. ____

____

____

5. ____

____

____

Name____ Date ____________________

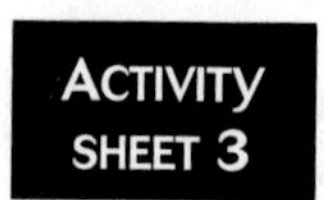

Surveying and Sampling

In the space below, write the description of the topic assigned to your team.

____

____

____

____

With your teammates, write a survey item based on this topic. Use the space below.

____

____

____

____

____

____

____

____

____

____

____

____

Surveying and Sampling

TEAM NUMBER 1

With your teammates, determine if the following method of collecting data would work well in your school. Remember, you want to generate a random sample. Your response will be written on chart paper and shared with the entire class.

Sampling Method 1:
Survey all students with a last name beginning with a vowel.

- -

Surveying and Sampling

TEAM NUMBER 2

With your teammates, determine if the following method of collecting data would work well in your school. Remember, you want to generate a random sample. Your response will be written on chart paper and shared with the entire class.

Sampling Method 2:
Survey the first 100 students who enter the school building in the morning.

- -

Surveying and Sampling

TEAM NUMBER 3

With your teammates, determine if the following method of collecting data would work well in your school. Remember, you want to generate a random sample. Your response will be written on chart paper and shared with the entire class.

Sampling Method 3:
Survey all students in English class second period.

- -

(continued)

Surveying and Sampling

Team Number 4

With your teammates, determine if the following method of collecting data would work well in your school. Remember, you want to generate a random sample. Your response will be written on chart paper and shared with the entire class.

Sampling Method 4:
Survey every 10th student who enters the cafeteria at lunch.

Surveying and Sampling

Team Number 5

With your teammates, determine if the following method of collecting data would work well in your school. Remember, you want to generate a random sample. Your response will be written on chart paper and shared with the entire class.

Sampling Method 5:
Using an alphabetical list of students, survey every 10th one.

Surveying and Sampling

Team Number 6

With your teammates, determine if the following method of collecting data would work well in your school. Remember, you want to generate a random sample. Your response will be written on chart paper and shared with the entire class.

Sampling Method 5:
Survey all students with blue eyes.

Name_______________________________ Date _______________

Surveying and Sampling

Write a description of what you have learned about surveying and sampling. Include:

- the types of questions that can be translated into the "item" format
- the various types of response formats
- randomness

Be sure to include in your writing:

- an analysis of data collection methods that are, and are not, random in nature

You may use the lines below and the other side of this sheet. Or you may use a separate sheet of paper. Your work will be assessed according to the criteria listed on the following page.

Name ________________________________ Date ________________

Surveying and Sampling

You have been required to submit a paper describing what you have learned about making surveys and what you have learned about sampling. Your paper will be assessed according to the following criteria and rubrics:

PRESENTATION

0—No paper submitted.
1—Paper submitted.
2—Paper is somewhat neat *or* organized.
3—Paper is neat *or* organized.
4—Paper is neat *and* organized.

SURVEY CONTENT

0—No reference to surveying.
1—Paper references surveying.
2—Paper demonstrates some achievement towards expected learning.
3—Paper demonstrates expected achievement in both types of questions and response formats.
4—Paper demonstrates achievement beyond expectations.

SAMPLING CONTENT

0—No mention of sampling.
1—Paper references sampling.
2—Paper demonstrates some achievement of expected learning.
3—Paper demonstrates expected achievement in both sampling methods and randomness.
4—Paper demonstrates achievement beyond expectations.

Average score (nearest 10th) = ________

GRADING TRANSFER

POINT AVERAGE PER OUTCOME:

From (not including) to	Letter grade	From (not including) to	Letter grade
3.1–4.0	A	1.9–2.4	C
3.0–3.1	A–	1.8–1.9	C–
2.9–3.0	B+	1.7–1.8	D+
2.6–2.9	B	1.2–1.7	D
2.5–2.6	B–	1.1–1.2	D–
2.4–2.5	C+	1.1 or below	F

Comments: ____
____
____

Graph-a-Bell

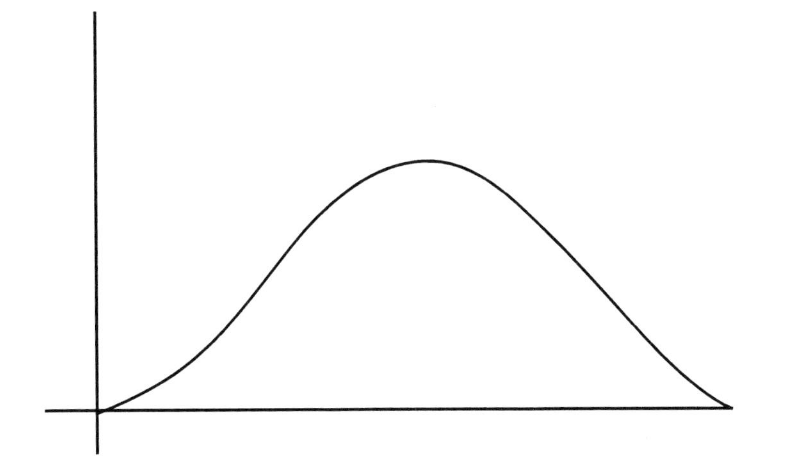

Teacher Page

Graph-a-Bell

Learning Outcome

Students will demonstrate:

- an understanding of the bell curve.

Time Requirements

4 hours

Materials

- Reproducible student handouts
- Calculators
- Graph paper
- Red and blue pens

NCTM Standards (2000) Addressed

Data Analysis and Probability:

- Formulate questions that can be addressed with data, and collect, organize, and display relevant data to answer them
- Select and use appropriate statistical methods to analyze data

Prerequisites

Students should have a working understanding of a frequency distribution chart.

Overview

This series of activities explores the concept of a bell curve and the terms associated with a bell curve, deviation and standard deviation.

Assessment

Distribute the Assessment Criteria sheet that follows Activity Sheet 4, and explain to students that their work will be assessed according to these criteria. Students should then attach the Assessment Criteria sheet to their work when they hand it in.

Procedures

Part One

Suggested time: 30 minutes

1. Divide class into pairs. Distribute Activity Sheet 1, and clarify directions. Students should brainstorm with their partners, then write individual stories on the back of their activity sheets.

(continued)

2. Assess student responses to determine which concepts are understood, which are partially understood, and which are already a part of the student's knowledge. Use a narrative form of feedback when providing assessment.

Part Two

Suggested time: 1 hour 15 minutes

3. Divide class into pairs. Distribute Activity Sheet 2 and graph paper. Direct students to transfer the data from the frequency distribution chart to a graph. At first, no directions should be given about how to create the graph.

4. After students have completed their work, use a reverse jigsaw approach to enhance student understanding. Randomly create teams of six students. Students convene in their new teams to share what they have done.

5. After the sharing session, assign students to individual seating. Instruct students to write a reflection about what they saw in common among the various graphical displays.

6. Assess student work. Narrative is most helpful to students during this assessment cycle. If the assessment indicates that students are, generally, unable to form bell curves, conduct a class discussion in order to lead students to the desired result. Use direct teaching—expository methods—if necessary. When students are assessed to be near the desired outcome, proceed to the next part of the activity.

Part Three

Suggested time: 30 minutes

7. Distribute graph paper and Activity Sheet 3. Review and clarify the directions. Students should complete the activity individually.

8. Collect, assess, and score student work using the assessment criteria established with Activity Sheet 2. Offer narrative feedback on the students' progress toward the expected outcomes.

Part Four

Suggested time: 45 minutes

9. Divide class into pairs. Be sure that each student has his or her own data and bell curves from the previous activities.

(continued)

10. Distribute Activity Sheet 4. Explain and clarify the directions. The use of calculators to determine percentages is highly recommended. Students are urged to edit their original worksheets and prepare them for a grading assessment. Use the Assessment Criteria sheet following Activity Sheet 4 to assess and score student work.

Part Five

Suggested time: 30 minutes

11. Divide the class into pairs and distribute Activity Sheet 5. Explain and clarify the directions. Since the students will be building on their knowledge, assessment for this activity sheet is informal. Feedback similar to that for Activity Sheet 1 is encouraged.

12. After students complete Activity Sheet 5, use a reverse jigsaw approach to enhance student responses. Randomly create teams of six. Students convene in their new teams to share what they have done. After the sharing session, instruct students to return to their partner on Activity Sheet 5 and share their findings. Encourage students to edit their results. Using the results of this activity sheet, students will complete the next activity.

Part Six

Suggested time: 30 minutes

13. Students will complete this part of the activity individually. Students are encouraged to use the results of the previous activity to assist them in responding. Distribute Activity Sheet 6. Explain and clarify the directions and the assessment criteria.

Part Seven

14. At this point, students should have a strong enough background to be formally introduced to the terms *deviation* and *standard deviation*. Distribute Activity Sheet 7, and have students complete it to assist in their understanding of the terms.

Name____ Date ____________________

Graph-a-Bell

Partner's Name ____________________________

Based on the graph below, make up a story that could explain the data. Each partner will write his or her own version of story.

The Abscissa axis (x-axis) does not represent time units.

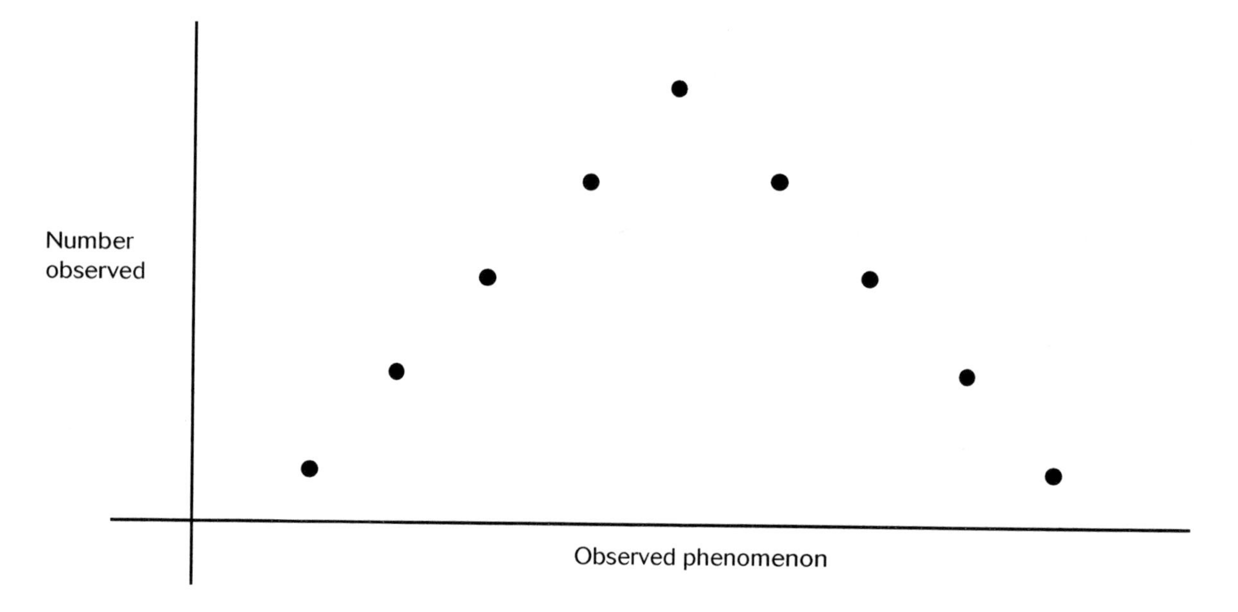

On the back of this sheet, write the story that you have created. Your partner will also write a version of the story on his or her activity sheet.

Name_______________________________ Date _______________

Graph-a-Bell

Recently 1,000 high school freshmen were surveyed about their television viewing habits. One survey item measured the number of hours per week that students watched television. The following Frequency Distribution Chart displays the results.

NUMBER OF HOURS FROM–TO (NOT INCLUDING)	NUMBER OF RESPONSES
0–3	2
3–6	11
6–9	24
9–12	49
12–15	79
15–18	123
18–21	138
21–24	150
24–27	139
27–30	117
30–33	77
33–36	43
36–39	26
39–42	15
42 or more	2

With your partner, create a graphical display of the data. Use the graph paper provided by your teacher. You will be assessed according to the criteria listed below.

PRESENTATION

0—No work submitted.
1—Work submitted.
2—Work is somewhat neat *or* somewhat accurate.
3—Work is neat *or* accurate.
4—Work is neat *and* accurate.

ACCURACY

0—No work submitted.
1—Work submitted.
2—Work somewhat accurate.
3—Work demonstrates total understanding of the methods of forming a bell curve.
4—Work demonstrates knowledge of graphical displays beyond expectations.

Name____ Date ____________________

Graph-a-Bell

Using your work from Graph-a-Bell, Activity Sheet 2, complete the following. Your work will be assessed according to the criteria listed on Activity Sheet 2.

1. On the grid below, redraw:

   - the axes

   - the data points

2. Draw a smooth curve through these data points.

SCORES

Graph ________ Presentation ________

Name________________________________ Date ________________

Graph-a-Bell

This is an assessment activity. You may need to edit previous work before you submit it.

Use the television survey data from the previous activities. The mean of the data (rounded to the nearest 10th) is 22.1. It is the focal point for this activity.

Compute the percentages of data that fall within the parameters identified.

NUMBER OF HOURS	PERCENT OF DATA WITHIN RANGE	NUMBER OF HOURS	PERCENT OF DATA WITHIN RANGE
21–24	________	9–36	________
18–27	________	6–39	________
15–30	________	3–42	________
12–33	________	0–42+	________

1. Redraw your bell curve, if necessary, to make it as accurate and as neat as possible.
2. On the bell curve, in *blue* ink, draw vertical lines so that $\frac{2}{3}$ of the data around the mean falls between the lines.
3. On the bell curve, in *red* ink, draw two more vertical lines so that 95% of the data falls between the lines.

Name________________________________ Date ________________

Graph-a-Bell

PRESENTATION

0—No work submitted.
1—Work submitted.
2—Work is somewhat neat *or* organized.
3—Work is neat *or* organized.
4—Work is neat *and* organized.

THE GRAPH

0—No work submitted.
1—Work submitted.
2—Work demonstrates some understanding of how to construct a graph.
3—Work demonstrates expected achievement towards the construction of a bell curve.
4—Work demonstrates a sophisticated and creative ability to create graphical displays.

THE BELL CURVE

0—No work submitted.
1—Work submitted.
2—Work demonstrates some understanding of the aspects of a bell curve.
3—Work demonstrates that the student is able to interpret a bell curve for percentages of responses, "smoothness" of the curve, and accuracy of statistical interpretation.
4—Work demonstrates an ability beyond expectations in creating, understanding, or interpreting a bell curve.

GRADING TRANSFER

POINT AVERAGE PER OUTCOME:

From (not including) to	Letter grade	From (not including) to	Letter grade
3.1–4.0	A	1.9–2.4	C
3.0–3.1	A–	1.8–1.9	C–
2.9–3.0	B+	1.7–1.8	D+
2.6–2.9	B	1.2–1.7	D
2.5–2.6	B–	1.1–1.2	D–
2.4–2.5	C+	1.1 or below	F

Comments: ____

____

____

____

Name______________________ Date ______________

Graph-a-Bell

With your partner, determine responses to the prompts below. Use a pencil.

1. Each of the graphs below represents a bell curve shape. Draw a vertical line through the middle of the curve. This line will approximate the mean of the graphed data.

2. Keep the mean line you just drew in the center. Draw two lines, equidistant from the mean line, within which $\frac{2}{3}$ of the data would lie.

3. Keep the mean line in the center. Draw two more vertical lines within which 95% of the data would lie.

Share your results with other students from other teams.

A.

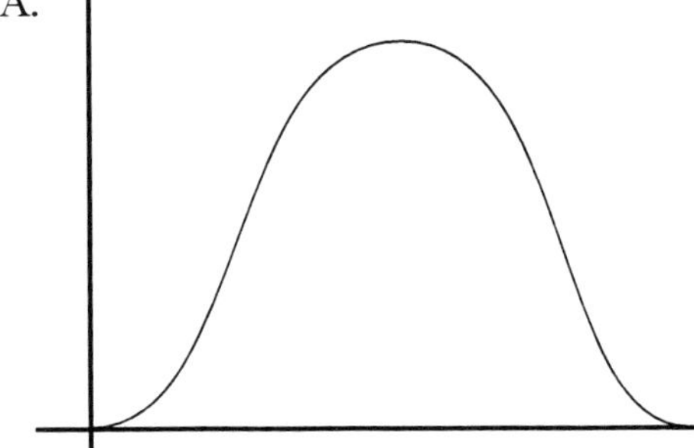

C.

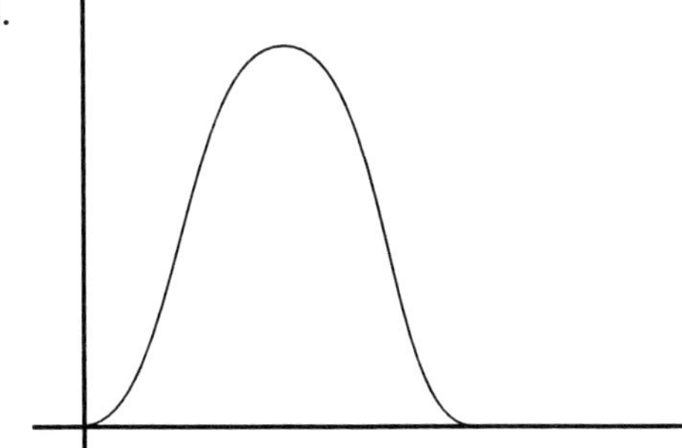

B.

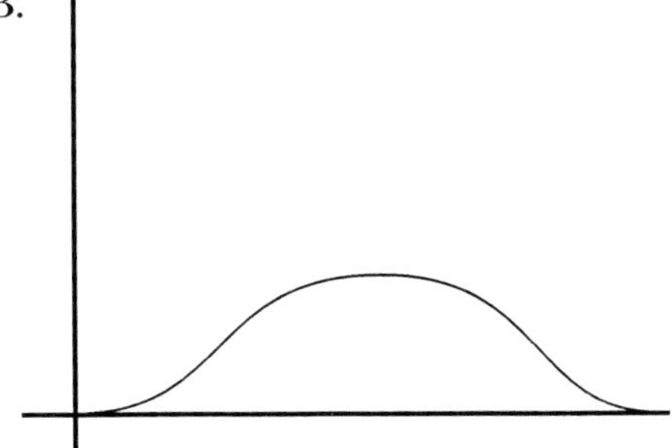

D.

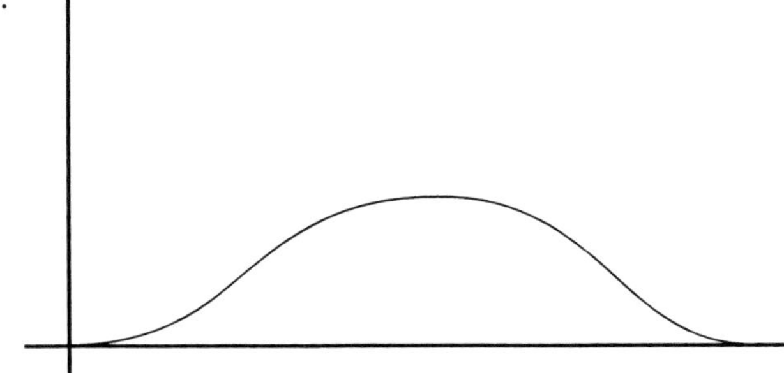

Name____ Date ____________________

Graph-a-Bell

In the space below, carefully sketch three different shapes of bell curves. On each of the curves, draw the vertical lines that would represent:

- the mean
- the area in which $\frac{2}{3}$ of the data would lie around the mean line
- the area in which 99% of the data would lie around the mean line

Your work will be assessed according to the criteria established earlier in this lesson.

SCORES

Presentation __________ Average __________
Bell curve __________ Grade __________
Graph __________

Name____ Date ____________________

Graph-a-Bell

Compute the mean of the following data set:

{4, 12, 4, 3, 4, 8, 7, 8, 5, 5, 6, 1, 9, 8}

Mean = ________

Write the mean in the first column below. In the second column, write each of the data points in order. Compute the answer as indicated.

MEAN –	DATA POINTS	= ANSWER	SQUARE EACH ANSWER

(continued)

Name____ Date ____________________

Graph-a-Bell *(continued)*

Directions: Using the chart on the previous page, follow the steps outlined below.

Compute the sum of the numbers in the "square" column = ____________

Then, let Q = the sum shown above.

Let N = the number of data.

Q/(N–1) = _____ (This is called the **variance**.)

Now find the square root of the variance.

This answer is called the **standard deviation.**

The theory of statistics declares that at least ⅔ of a data set will lie within one standard deviation of the mean.

The theory of statistics declares that at least 95% of a data set will lie within two standard deviations of the mean.

Compute: mean + standard deviation = ________________

Compute: mean – standard deviation = ________________

What percent of the data occur between these two numbers? ________

Compute: mean + 2 standard deviations = ________________

Compute: mean – 2 standard deviations = ________________

What percent of the data occur between these two numbers? ________

Are these findings consistent with the theory? Why or why not? Use the back of this sheet for your answer.

Share Your Bright Ideas with Us!

We want to hear from you! Your valuable comments and suggestions will help us meet your current and future classroom needs.

Your name ____________________ Date ____________

School name ____________________ Phone ____________

School address ____________________

Grade level taught ______ Subject area(s) taught ____________ Average class size ____

Where did you purchase this publication? ____________________

Was your salesperson knowledgeable about this product? Yes ____ No ____

What monies were used to purchase this product?

___School supplemental budget ___Federal/state funding ___Personal

Please "grade" this Walch publication according to the following criteria:

Quality of service you received when purchasing	A	B	C	D	F
Ease of use	A	B	C	D	F
Quality of content	A	B	C	D	F
Page layout	A	B	C	D	F
Organization of material	A	B	C	D	F
Suitability for grade level	A	B	C	D	F
Instructional value	A	B	C	D	F

COMMENTS: ____________________

____

What specific supplemental materials would help you meet your current—or future—instructional needs?

____

Have you used other Walch publications? If so, which ones? ____________________

May we use your comments in upcoming communications? ___Yes ___No

Please **FAX** this completed form to **207-772-3105**, or mail it to:

Product Development, J.Weston Walch, Publisher, P.O. Box 658, Portland, ME 04104-0658

We will send you a **FREE GIFT** as our way of thanking you for your feedback. **THANK YOU!**